Transit Bus Applications of Lithium Ion Batteries: Progress and Prospects

DECEMBER 2012
FTA Report No. 0024

PREPARED BY

Dr. Aviva Brecher
Energy Analysis and Sustainability Division
Energy and Environmental Systems Technical Center
Volpe National Transportation Research Center

SPONSORED BY

Federal Transit Administration
Office of Research, Demonstration and Innovation
U.S. Department of Transportation
1200 New Jersey Avenue, SE
Washington, DC 20590

AVAILABLE ONLINE

http://www.fta.dot.gov/research

Metric Conversion Table

SYMBOL	WHEN YOU KNOW	MULTIPLY BY	TO FIND	SYMBOL
\multicolumn{5}{c}{LENGTH}				
in	inches	25.4	millimeters	mm
ft	feet	0.305	meters	m
yd	yards	0.914	meters	m
mi	miles	1.61	kilometers	km
		VOLUME		
fl oz	fluid ounces	29.57	milliliters	mL
gal	gallons	3.785	liters	L
ft^3	cubic feet	0.028	cubic meters	m^3
yd^3	cubic yards	0.765	cubic meters	m^3

NOTE: volumes greater than 1000 L shall be shown in m^3

SYMBOL	WHEN YOU KNOW	MULTIPLY BY	TO FIND	SYMBOL
		MASS		
oz	ounces	28.35	grams	g
lb	pounds	0.454	kilograms	kg
T	short tons (2000 lb)	0.907	megagrams (or "metric ton")	Mg (or "t")
		TEMPERATURE (exact degrees)		
°F	Fahrenheit	5 (F-32)/9 or (F-32)/1.8	Celsius	°C

Selected SI Units and Equivalents for Batteries

Unit	Equivalent	Measured	Quantity
1 Kilowatt (kW)	1,000	Watts (W)	Power
1 Watt	1	Joule/second (J/s)	Power
1 kW-hour (kWh)	3.6	MegaJoules (MJ)	Energy
1 Watt-second	1	Joule	Energy
1 GigaJoule (GJ)	277.8	kWh	Energy
1 kWh/ Kilogram (kg)	3.6	GigaJoules/ton	Energy density
1 Coulomb (C)	1	Ampere-second (As)	Electric charge
1 AmpHour (Ah)	3600	Coulomb	Electric charge
1 Ampere (A)	1	Coulomb/second (C/s)	Electric current
1 Volt (V)	1	Joule/Coulomb (J/C)	Electric Voltage
1 Farad (F)	1	Amp-second/Volt	Capacitance

Source: DOE Energy Information Administration (EIA) and www.mpoweruk/conversion_table.htm

REPORT DOCUMENTATION PAGE	Form Approved OMB No. 0704-0188

Public reporting burden for this collection of information is estimated to average 1 hour per response, including the time for reviewing instructions, searching existing data sources, gathering and maintaining the data needed, and completing and reviewing the collection of information. Send comments regarding this burden estimate or any other aspect of this collection of information, including suggestions for reducing this burden, to Washington Headquarters Services, Directorate for Information Operations and Reports, 1215 Jefferson Davis Highway, Suite 1204, Arlington, VA 22202-4302, and to the Office of Management and Budget, Paperwork Reduction Project (0704-0188), Washington, DC 20503.

1. AGENCY USE ONLY	2. REPORT DATE December 2012	3. REPORT TYPE AND DATES COVERED 2007-2012
4. TITLE AND SUBTITLE Transit Bus Applications of Lithium Ion Batteries: Progress and Prospects		5. FUNDING NUMBERS MA-26-7200
6. AUTHOR(S) Dr. Aviva Brecher, Principal Technical Advisor, Energy Analysis and Sustainability Division, Energy and Environmental Systems Technical Center		
7. PERFORMING ORGANIZATION NAME(S) AND ADDRESSE(ES) DOT/RITA Volpe National Transportation Systems Center 55 Broadway Cambridge, MA 021542 www.volpe.dot.gov		8. PERFORMING ORGANIZATION REPORT NUMBER FTA Report No. 0024
9. SPONSORING/MONITORING AGENCY NAME(S) AND ADDRESS(ES) U.S. Department of Transportation Federal Transit Administration Research, Demonstration and Innovation East Building 1200 New Jersey Avenue, SE Washington, DC 20590		10. SPONSORING/MONITORING AGENCY REPORT NUMBER FTA Report No. 0024
11. SUPPLEMENTARY NOTES Available Online [http://www.fta.dot.gov/research]		
12A. DISTRIBUTION/AVAILABILITY STATEMENT Available from: National Technical Information Service (NTIS), Springfield, VA 22161. Phone 703.605.6000, Fax 703.605.6900, email [orders@ntis.gov]		12B. DISTRIBUTION CODE TRI-20

13. ABSTRACT

This report provides an overview of diverse transit bus applications of advanced Lithium Ion Batteries (LIBs). The report highlights and illustrates several FTA programs that fostered the successful development, demonstration, and deployment of fuel-efficient hybrid-electric and electric drive transit buses in operational urban fleets over the last decade. The focus is on recent progress in the rechargeable energy storage systems (RESS) that successfully integrated the lighter, more compact LIBs with higher energy density and capacity in a broad range of power and propulsion configurations for urban transit bus fleets. Improvements in fuel efficiency and environmental performance of succeeding generations as well as LIB-related safety, cost, reliability, availability, and maintainability challenges are discussed in context, including recent recalls due to LIB safety issues. Progress in and prospects for future LIB improvements and remaining bus application challenges are also discussed.

14. SUBJECT TERMS Hybrid-electric bus (HEB); electric bus (EB); fuel cell bus (FCB); Federal Transit Administration (FTA); National Fuel Cell Bus Program (NFCBP); Transit Investments in Greenhouse Gas and Energy Reduction (TIGGER); Lithium Ion Battery (LIB); Rechargeable Energy Storage System (RESS); Auxiliary Power Unit (APU)		15. NUMBER OF PAGES 42	
16. PRICE CODE			
17. SECURITY CLASSIFICATION OF REPORT Unclassified	18. SECURITY CLASSIFICATION OF THIS PAGE Unclassified	19. SECURITY CLASSIFICATION OF ABSTRACT Unclassified	20. LIMITATION OF ABSTRACT

TABLE OF CONTENTS

1	Executive Summary
3	Section 1: Introduction
3	Background and Scope
3	Deployment Trends for Electric Drive in Transit Buses
7	Section 2: Integration of Lithium Ion Batteries in Electric Drive Buses
10	Section 3: Examples of Hybrid and Battery Electric Transit Buses with LIB-Based RESS
10	Overview of Transit Buses with LIBs
11	BAE Systems HybriDrive with Lithium Iron Phosphate (LFP) Battery
12	Proterra Bus TerraVolt RESS with Lithium Titanate Battery
13	ISE Corporation Thunderpack Integration of LIBS with Ultracapacitor APU
15	Designline EcoSaver IV HEB with LIB and Capstone Microturbine APU
16	Enova Systems HybridPower for School and City PHEBs
17	FTA Advanced Transit Bus Demonstration and Deployment Programs
17	National Fuel Cell Bus Program (NFCBP)
20	FTA TIGGER and Clean Fuels Grant Programs
22	Section 4: Lessons Learned, Progress, and Prospects
22	Case Studies and Safety Lessons Learned from LIB Bus Operations
25	LIB Bus Market Prospects and Challenges
30	Acronyms

LIST OF FIGURES

8	Figure 2-1:	Schematic of BAE Systems HybriDrive propulsion system used in Daimler Orion VII and New Flyer Xcelsior hybrid-electric buses
8	Figure 2-2:	BAE rechargeable energy storage system (RESS) in Orion VII using Lithium Ion Iron Nano-Phosphate Batteries from A123 Systems
9	Figure 2-3:	Proterra BE35 EcoRide Battery Electric Bus (BEB)
9	Figure 2-4:	Battery used in Proterra buses uses 50 Ah Lithium Ion Nano-Titanate cells produced by Altairnano
14	Figure 3-1:	ISE Hybrid Drive integrated in fuel cell hybrid bus
15	Figure 3-2:	Lithium Ion Battery modules in BLUWAYS ISE energy storage system, with master control module attached
16	Figure 3-3:	Charm City Circulator, a DesignLine EcoSaver IV operating in Downtown Baltimore
18	Figure 3-4:	SunLine Fuel Cell Bus

ACKNOWLEDGMENTS

The principal author of this report is Dr. Aviva Brecher, Principal Technical Advisor on Energy Analysis and Sustainability in the Energy and Environmental Systems Technical Center at the U.S. DOT Volpe National Transportation Systems Center (Volpe Center). The author gratefully acknowledges the guidance and sponsorship of Matthew Lesh, Transportation Program Specialist, in the Office of Mobility Innovation Research, Demonstration & Innovation, Federal Transit Administration. Volpe Center colleague Stephen Costa of the Energy Analysis and Sustainability Division is thanked for valuable technical assistance; Cynthia Sabin and Eva Dykstra are thanked for editorial help. Volpe Center Management is acknowledged for supporting the author's efforts to prepare a chapter based on this report for publication in the 2012 Elsevier Science and Technology Handbook of Lithium Ion Batteries Applications.

ABSTRACT

This report provides an overview of the diverse transit bus applications of advanced Lithium Ion Batteries (LIBs). The focus is on recent progress in the rechargeable energy storage systems (RESS) that successfully integrated the lighter, more compact LIBs with higher energy density and capacity in a broad range of power and propulsion configurations for urban transit bus fleets. The report highlights and illustrates several Federal Transit Administration (FTA) grant programs that have fostered the successful development, demonstration, and deployment of fuel-efficient hybrid-electric and electric drive transit buses in operational urban fleets over the last decade. Improvements in fuel efficiency and environmental performance of succeeding generations as well as LIB-related safety, cost, reliability, availability, and maintainability challenges are discussed in context. Progress and prospects for future LIB improvements and remaining bus application challenges are also discussed.

EXECUTIVE SUMMARY

Elsevier Science & Technology invited the author to prepare a chapter highlighting recent transit bus applications of advanced Lithium Ion Batteries (LIBs) for its 2012 *Handbook of Lithium Ion Batteries Applications*. The chapter will highlight successful federal programs that have developed, demonstrated, and deployed advanced, fuel-efficient, and environmentally-friendly transit buses to inform a broadly interdisciplinary international technical audience. The report reviews the diverse LIB-based energy storage systems currently deployed in operational hybrid, electric, and fuel cell transit buses and is intended as an open source, Web-postable version of the handbook chapter for public transit stakeholders.

Section 1 provides the background and rationale for upgrading urban transit bus fleets with advanced LIBs to improve their energy efficiency and environmental performance. Transit bus fleets are recognized as a promising technology demonstration platform and an early adopter market niche for deploying advanced technologies, such as the hybrid-electric drive trains with rechargeable energy storage systems (RESS). Public transit buses also provide a good test-bed for evaluating new battery and hybrid-electric technologies, given their operation on pre-determined routes and predictable duty cycle and schedules, benefitting from professional drivers and regular preventive maintenance or repair in large depots. As fuel prices rise and stricter environmental regulations are enforced, the diesel urban transit bus fleet in the U.S. has become cleaner and "greener" through the adoption of alternative fuels (natural gas, biodiesel) and/or advanced power trains, such as hybrid-electric, electric, and fuel cell auxiliary power units refueled by hydrogen.

Commercially-available hybrid-electric buses (HEBs), electric buses (EBs), and fuel-cell buses (FCBs) provide 30–50 percent better fuel efficiency in city operation over their diesel counterparts, albeit at higher initial cost by factors of 2–3, but promising paybacks in lifecycle costs. The recent use of LIBs in hybrid and electric bus transit fleets has been facilitated by their centralized operations, trained staff, scheduled maintenance, and multi-year warranties; however, issues such as electrical system safety remain a concern. Since LIB technology is relatively new and rapidly evolving, the longest warranty coverage of a current bus equipped with LIB is 5–6 years—only half of the 12-year minimum required service life of transit buses.

Section 2 discusses proven options for integrating LIBs in RESS subsystems used in advanced transit buses. Selected illustrations of the diverse hybrid and electric drive-trains, LIB chemistries, and RESS configurations adopted in operational hybrid and electric drive buses are provided. The RESS on board the bus is designed to efficiently capture, store, and deliver regenerated braking energy for reuse. Both RESS architecture and requirements for energy capacity and power delivered at steady state or peak load depend on the bus route, range, and duty cycle.

Section 3 details specific examples of commercial transit buses, which illustrate the successful integration of LIBs into electric drivetrains for several operational and emerging U.S. bus applications. In the past five years, hybrid and electric propulsion drivetrains were upgraded with the lighter, smaller, and more powerful LIBs, replacing the earlier Nickel Metal Hydride (NIMH) and lead acid (PbA) heavy-duty batteries. LIBs are controlled by advanced power electronics, Battery Management Systems (BMS), and Thermal Management Systems (TMS) and are sometimes enhanced with an Auxiliary Power System (APU). The APUs used to complement the battery and extend its range may consist of micro-turbine generators, fuel cells (FC), or ultracapacitors (ucaps). External RESS charging infrastructures for slow or fast recharge and range extension of plug-in hybrids and all electric buses have also been developed and deployed.

Diverse LIB bus applications to hybrid, electric, and fuel cell buses are illustrated with U.S. operational examples of HEBs, EBs, and FCBs that integrate LIBs within diverse RESS and powertrain configurations. Major Federal Transit Administration (FTA) research and technology (R&T) and grant programs enabling public transit fleet deployments are highlighted. Advanced electric-drive bus technology and LIB commercial viability advances are illustrated for representative competitive grant programs, such as the National Fuel Cell Bus Program (NFCBP), Clean Fuels grants, and the Transit Investments for Greenhouse Gas and Energy Reduction (TIGGER) program.

Section 4 summarizes key lessons learned to date, focusing on case studies of LIB operational safety, remaining challenges for large-scale deployment of LIBs in urban bus fleets, and a brief assessment of progress and prospects. Other lessons learned regard LIB performance, reliability, durability, and cost, which remain the primary challenges for large-scale deployment in public transit fleets and other bus fleet applications (e.g., school buses, airport shuttle buses, etc.). Based on rapid progress in the state of the art for LIB chemistry, energy capacity, energy density, and life-cycle performance, as well as cost reduction efforts underway, accelerated LIB deployment in bus fleets and other heavy-duty applications is anticipated.

SECTION 1

Introduction

Background and Scope

Elsevier Science & Technology invited the author to prepare a chapter on bus applications of LIBs for its 2012 Handbook of Lithium Ion Batteries based on the author's earlier report for the Federal Transit Administration (FTA) that reviewed the state of the art of energy storage systems onboard transit buses and remaining research and technology gaps.[1] The chapter needed to summarize and update that previous FTA report and complement other chapters that address novel Lithium Ion Battery (LIB) chemistries, performance, and cost trends and challenges as well as LIB applications to vehicles and utilities. FTA and the U.S. Department of Transportation (DOT)'s Research and Innovative Technology Administration (RITA) and Volpe National Transportation Systems Center co-sponsored this effort to make the material available free of charge to the transit community. Since the handbook is intended to broadly inform a diverse international science and technology readership, this report (to be converted into a chapter) focuses on introducing FTA and related U.S. federal programs that promote the development, demonstration, integration, and deployment of LIBs in advanced transit buses. Section 2 provides a snapshot of the diverse LIB chemistries and configuration options for their integration into hybrid and electric drive buses as well as recent market trends. Section 3 illustrates the successful integration of LIBs into electric drivetrains for several operational and emerging U.S. bus applications. Section 4 summarizes the key lessons learned, remaining challenges for large-scale deployment of LIBs in urban bus fleets, and a brief assessment of progress and prospects.

Deployment Trends for Electric Drives in Transit Buses

In 2010, there were about 65,000 public transit buses and 850,000 commercial and school buses in the U.S. bus fleet. Statistics from the American Public Transportation Association (APTA)[2] indicate that 33 percent of U.S. public transit buses use alternative fuels, including 7 percent hybrid buses and 0.1 percent electric buses. Urban transit bus fleets comprise a very visible and well-accepted platform for demonstrating and implementing green technologies and alternative

---

[1] A. Brecher, "Assessment of Needs and Research Roadmaps for Rechargeable Energy Storage Systems (RESS) Onboard Electric Drive Buses," FTA-TRI-MA-26-7125-2011.1, December 2010. Available at http://ntl.bts.gov/lib/35000/35700/35796/DOT-VNTSC-FTA-11-01.pdf.

[2] 2011 APTA Public Transportation Factbook, http://www.apta.com/resources/statistics/Documents/FactBook/APTA_2011_Fact_Book.pdf.

fuels in metro areas, which must reduce ambient air pollution and energy consumption. Bus fleets are also recognized as a promising market niche[3] for the manufacturers, suppliers, and integrators of advanced energy recovery and storage technologies and of fuel efficiency products. The early adoption of innovative hybridization and advanced rechargeable energy storage systems (RESS) and battery technologies in urban transit bus fleets is facilitated by structured (fixed-route and schedule) transit operations and by centralized maintenance. Transit bus fleets are also managed, operated, and maintained by trained professional staff who ensure and enforce system safety best practices.

Over the past two decades, compliance with stricter environmental requirements led to the rapid penetration of HEBs and EBs in urban transit bus fleets to improve fuel efficiency by 20–50 percent and, correspondingly, reduce pollutant emissions. Advanced technologies (lightweight materials for body, chassis, and seat assemblies; stop-start systems for idle reduction; improved batteries, electric motors, converters, and power electronics) are also being deployed to further improve the fuel efficiency of advanced buses and urban air quality.

Transit buses operate on a predictable route, thus allowing for the optimization of RESS and associated Power Management Systems (PMS) to match specific route, range, duty cycle, and environmental conditions. The continuous operation and stop-start duty cycle of urban buses imposes stricter reliability, availability, maintainability, and durability requirements on RESS than for light-duty vehicles, although mass, size, and cost requirements are relatively less severe. LIB integration of bus fleets over the past decade has offered greater performance (energy storage capacity and density) at lower power-pack volume and mass but at a cost premium.

Transit bus competitive grant programs funded by FTA in recent years—such as Bus and Bus Facilities, Clean Fuels, and Transit Investments for Greenhouse Gas and Energy Reduction (TIGGER)[4]—enabled the demonstration and rapid adoption of electric and hybrid drive and advanced batteries in hybrid-electric buses (HEBs), electric buses (EBs), and fuel-cell buses (FCBs) for urban fleets and/or fleet expansion and renewal.

---

[3] "The Transit Bus Niche Market for Alternative Fuels: Module 8—Overview of Advanced Hybrid and Fuel Cell Bus Technologies," DOE Clean Cities Coordinator Toolkit, TIAX LLC, December 2003. Available at www.afdc.energy.gov/pdfs/mod08.zebs.pdf; Goldman Sachs, "Energy Storage: Advanced Batteries," 2010, http://www.eosenergystorage.com/articles/GSBatteryReport2010-06-29.pdf; M. Lowe, "Lithium Ion Batteries for Electric Vehicles: The U.S. Value Chain," 2010. Available at http://unstats.un.org/unsd/trade/s_geneva2011/refdocs/RDs/Lithium-Ion%20Batteries%20(Gereffi%20-%20May%202010).pdf.

[4] See FTA Formula and Discretionary Grant programs, http://www.fta.dot.gov/grants_263.html.

Public transit and school bus fleet adoption of fuel efficiency hybrid and electric technologies are also supported by the U. S. Department of Energy (DOE) Clean Cities program,[5] as well as by the American Recovery and Reinvestment Act (ARRA), with more than $300 million invested in hybrid and clean buses. The DOE Vehicle Technologies Program's advanced vehicle testing activities has identified several advantages of transit bus fleets as early adopters of new energy storage and fuel efficiency technologies:[6]

- Large transit fleets represent a substantial niche market.
- The average bus life of 12 years requires demonstrated high reliability and availability.
- Federal subsidies for renewal of transit bus fleets foster the adoption of alternative fuels and the deployment of innovative energy storage technologies.
- These technologies are commonly used in metro areas that need to curb air pollution and reduce energy consumption with high visibility and market impact.
- The operation of these technologies on fixed routes has predictable duty cycle, power loading, and energy requirements.
- These technologies are centrally fueled and maintained by trained technicians who are able to ensure safe operability.

The DOE National Renewable Energy Laboratory (NREL) has also assisted FTA by performing evaluations of advanced technology in service for diverse HEBs, EBs, and FCBs.[7]

The Environmental Protection Agency (EPA) has also facilitated the adoption of green diesel-electric hybrid school and transit buses through its Clean Diesel, Clean Bus, and SmartWay programs as well as recent ARRA awards.[8]

Since buses have an average service life of 12 years, they may be repowered with hybrid drive retrofit kits at a lower cost than the purchase of new HEBs. These deployments include use of lighter, smaller, and more powerful LIBs, replacing Nickel Metal Hydride (NiMH) batteries with integrated power and control electronics and any related charging infrastructure. Lighter-weight materials for body, chassis, seat assemblies, and other components have further improved the fuel efficiency of advanced electric drive buses and extended their operating range.

---

[5] See postings at http://www1.eere.energy.gov/cleancities/.

[6] See Transit Vehicles DOE/EERE postings at http://www.eere.energy.gov/topics/vehicles.html.

[7] See http://www.fta.dot.gov/documents/HydrogenandFuelCellTransitBusEvaluations42781-1.pdf, http://www.actransit.org/wp-content/uploads/NREL_rept_OCT2010.pdf, and http://www.fta.dot.gov/documents/HydrogenandFuelCellTransitBusEvaluations42781-1.pdf.

[8] See www.epa.gov/cleandiesel; www.epa.gov/recovery and http://www.epa.gov/smartway/financing/index.htm.

CALSTART, Inc.,[9] which has been engaged in partnerships to develop, demonstrate, and evaluate prototypes of advanced zero emission buses (ZEBs), has compiled for FTA a compendium of RESS battery chemistries for heavy-duty applications.[10] It reviewed the bus energy storage and traction power needs for buses and listed manufacturers, suppliers, and system integrators.

The best current RESS options integrate the lighter, smaller, and more powerful LIBs of diverse chemistries that have higher energy capacity (kilowatt-hour [kWh]) and peak power (kW) as well as higher energy density (in watt-hour [Wh] per kilogram [kg]) and power density (watt/kg) than NiMH and PbA batteries used in early hybrid and electric buses. The energy capacity for a hybrid bus battery pack ranges from 2–10 kWh, but more than 80 kWh is needed for an all-electric bus. The LIB energy density is in the range of 110–150 Wh/kg versus 40–140 Wh/kg for NiMH and 30–50 Wh/kg for best lead acid batteries. The power density for LIBs, essential for rapid acceleration, ranges from 200–3,000+ Wh/kg, depending on LIB chemistry, versus 200–1,300 for NIMH.

Over the past two decades, FTA has funded major research, technology, and pilot demonstration and deployments of innovative HEBs, EBs, and FCBs. Its multi-year Research Plan[11] and several competitive grant programs have enabled public transportation authorities to purchase and renew HEB/EB/FCB vehicle fleets and make major fuel efficiency improvements of bus facilities and associated charging infrastructure. The primary FTA competitive grant programs for clean buses integrating advanced LIB in diverse electric drive train architectures include the TIGGER,[12] Clean Fuels, and Bus and Bus Facilities programs.[13]

Industry and technical standards for LIB testing and for safety assurance are also being developed by the Society of Automotive Engineers (SAE)[14] and the National Fire Protection Association (NFPA) to facilitate the commercialization and large-scale adoption of LIB-based RESS in the nation's fleet of transit buses, school buses, and commercial coaches. These LIB-related safety issues are discussed in detail in Section 4. Prospects for expanded LIB integration in hybrid and electric buses and trucks are promising, provided that the remaining challenges in battery and system performance, safety, durability, and affordability can be overcome in the near future.

---

[9] See ZEB postings at www.calstart.org.

[10] See *Energy Storage Compendium: Batteries for Electric and. Hybrid Heavy Duty Vehicles*, March 2010, Available at www.calstart.org/news_and_publications/Publications.aspx.

[11] See *FTA Multi-Year Research Program Plan (FY 2009–2013)*, September 2008, http://www.fta.dot.gov/documents/FTA_TRI_Final_MYPP_FY09-13.pdf.

[12] See TIGGER program postings at http://www.fta.dot.gov/%20TIGGER.

[13] See FY12 FTA discretionary grant programs at http://www.fta.dot.gov/grants/13094.html and http://fta.dot.gov/documents/Discretionary_Webinar_Slides_for_FY_2012_-_2-29_and_3-1.pdf

[14] See "SAE Releases Li-ion Battery Safety Standards" at https://www.sae.org/mags/aei/SAEWC/9539 and NFPA March 2012 news releases at http://www.nfpa.org/newsReleaseDetails.asp?categoryid=488&itemId=55931&cookie_test=1.

SECTION 2
Integration of Lithium Ion Batteries in Electric Drive Buses

Transit buses operate on predictable routes, allowing for optimization of RESS and PMS to match range and duty cycle. However, the continuous operation and stop-start duty cycle of urban transit buses impose stricter reliability, availability, maintainability, and durability requirements than for light-duty vehicles. Over the past decade, NREL, in partnership with FTA, has conducted a series of evaluations of advanced hybrid-electric drive bus fleets that document the continuing improvements in RESS capabilities and performance for succeeding generations of vehicles.[15]

LIBs are integrated in the electric drive train with electric motors/generators, power conversion units (inverters), and power monitoring and control electronics to ensure safe and smooth bus operation. The PMS includes hardware and software for the Battery Management Systems (BMS) and Thermal Management Systems (TMS) that monitor and maintain a safe operating temperature to prevent degraded battery performance, as well as thermal runaway leading to fires or explosion.

State-of-the-art LIB chemistries and RESS technology options were reviewed in greater detail and evaluated to identify research needs in a recent FTA/Volpe Center report.[16] The chemistry, shape, size, energy density, and type of LIB integrated into the hybrid drive all depend on the electric drive hybrid configuration (series, parallel, or dual mode) and the duty cycle determining its depth of discharge: charge-sustaining for HEBs and FCBs, or charge-depleting for plug-in hybrid and electric buses. There are also multiple choices for the electric drive train configuration: series, parallel, split-parallel, post- and pre-transmission, engine-dominant or battery-dominant, plug-in hybrid, and pure electric. Each type of hybrid or electric drive train requires tailored RESS, optimized for bus duty cycle and route.

---

[15]See NREL Hybrid Electric Drive Fleet Test and Evaluation reports available at www.nrel.gov/vehiclesandfuels/fleettest/publications_hybrid.html.

[16]A. Brecher, "Assessment of Needs and Research Roadmaps for Rechargeable Energy Storage System (RESS) Onboard Electric Drive Buses," FTA-TRI-MA-26-7125-2011.1, December 2010. Available at http://ntl.bts.gov/lib/35000/35700/35796/DOT-VNTSC-FTA-11-01.pdf.

Figure 2-1 shows the BAE Systems Hybridrive configuration and placement of key power and propulsion subsystems for the widely-deployed Orion VII and the New Flyer Xcelsior HEBs, which integrated A123 lithium ion iron nano-phosphate (LiFePO4) LIBs, and Figure 2-2 shows the BAE RESS, which is placed on the roof for passive air cooling.

Figure 2-1

Schematic of BAE Systems HybriDrive propulsion system integrated in a transit bus, used in the Daimler Bus NA Orion VII and the New Flyer Xcelsior hybrid-electric buses

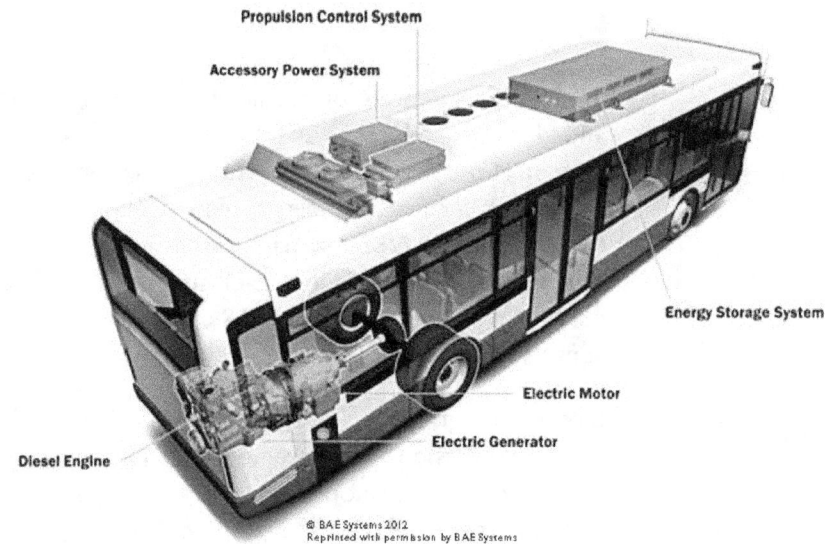

Illustration courtesy of BAE Systems Controls, Inc.

Figure 2-2

BAE rechargeable energy storage system (RESS) in Orion VII using Lithium Ion Iron Nano-Phosphate Batteries from A123 Systems

Illustration courtesy of BAE Systems Controls, Inc.

LIBs are often integrated into RESS with complementary Auxiliary Power Systems (APUs) to extend the bus range between recharges and to meet the duty cycle needs of enhanced energy capacity, peak power delivery, and storage of recaptured braking energy. The APUs may consist of FCs, micro-turbines (Capstone), ultracapacitors, and other types of batteries in a dual-battery bus. Complex drive train, energy storage, and PMS and cooling are also needed to accommodate these APUs. For example, FC stacks and compressed hydrogen storage tanks may be placed on the roof for easy access, facilitating maintenance and safe hydrogen venting in case of leakage.

The size and weight of batteries are not as critical for HEB/EB applications as are the energy capacity and power-delivery capabilities (the C-ratio). Buses can accommodate large-format batteries either under the floor, on the roof, or in the back of the bus for easy access. A typical large-format bus battery pack consists of several modules, each with multiple battery strings stacked in series or parallel, each string comprising hundreds to thousands of cells in array. A large-format LIB configuration with fewer strings in the stack and fewer connections is easier to monitor and to balance the voltage of individual cells. Cells may be of different shapes—arrays of prismatic cells are more compact than cylindrical cells and easier to stack in compact packaging. Figure 2-3 shows the composite-bodied electric Proterra EcoRide bus, which can be inductively recharged by overhead FastFill chargers at station stops; Figure 2-4 shows its TerraVolt modular LIB battery.

Figure 2-3
Proterra BE35 EcoRide Battery Electric Bus (BEB)

Photo courtsey of Proterra

Figure 2-4
Battery used in Proterra buses uses 50 Ah Lithium Ion Nano-Titanate cells produced by Altairnano

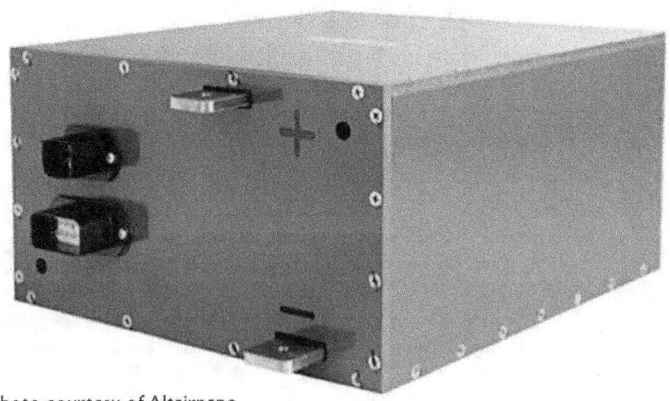

Photo courtesy of Altairnano

SECTION 3

Examples of Hybrid and Battery Electric Transit Buses with LIB-based RESS

Overview of Transit Buses with LIBs

The Volpe Center has assisted FTA in developing an Electric Drive Strategic Plan (EDSP) for research, technology, demonstration, and deployment as part of its multi-year research plan.[17] One EDSP activity was to evaluate onboard energy storage requirements for different types of electric drive transit buses with Lithium Ion and other battery and APU technology options.[18] The resulting report discussed a broad range of electric drive train configurations for HEBs/EBs/FCBs and the system integrators' rationale for the choice of LIB chemistries, battery size, weight, shape, and packaging as well as emplacement (safety, power and thermal management, energy density, rate of charge and discharge, durability, etc.).

The FTA-funded Transit Cooperative Research Program (TCRP) Report 132[19] provides useful guidance to transit authorities regarding the market readiness and lifecycle costs (LCC) of HEB/EB technologies, including some with LIBs. It found that the longevity and reliability of advanced batteries still falls short of transit agency expectations, and their capital and replacement costs are still high (up to $60K out of a bus cost that is 2–3 times higher than clean diesel or compressed natural gas [CNG] counterparts). TCRP Report 132 reviewed the lifecycle performance of commercially-available hybrid and electric buses with diverse power train configurations (series, parallel) and determined that series hybrids are preferred for city stop-and-go operation, while parallel hybrids are better for over-the-road route operations, with dual mode hybrids also available.

More complex power train and fuel storage configurations are needed for FCBs to accommodate both battery stacks and the APUs consisting of fuel cell stacks and

---

[17] See FTA Multi-year Research Program Plan (2009-2013) at http://www.fta.dot.gov/documents/FTA_TRI_Final_MYPP_FY09-13.pdf.

[18] A. Brecher, "Assessment of Needs and Research Roadmaps for Rechargeable Energy Storage System (RESS) Onboard Electric Drive Buses," 2010, FTA-TRI-MA-26-7125-2011.1, http://ntl.bts.gov/lib/35000/35700/35796/DOT-VNTSC-FTA-11-01.pdf.

[19] TRB/TCRP Report 132, "Assessment of Hybrid Electric Transit Bus Technology," December 2009, http://www.trb.org/Main/Blurbs/Assessment_of_HybridElectric_Transit_Bus_Technolog_162703.aspx.

the compressed hydrogen tanks fueling them. Pressurized hydrogen tanks or fuel tanks for onboard reforming are usually placed on the bus roof for ease of access for maintenance to facilitate safe venting of the flammable hydrogen in case of leakage.

Representative HEB/EB/FCBs using LIBs are selected here to illustrate LIB chemistry options and RESS integration architectures.

BAE Systems HybriDrive with Lithium Iron Phosphate (LFP) Battery

BAE Systems[20] is a major integrator and supplier of HybriDrive series hybrid-electric propulsion for Orion HEBs manufactured by Daimler Buses North America (DBNA).[21] HybriDrive was optimized for start-and-stop urban driving duty cycles and was improved over several generations of Daimler's Orion series of diesel hybrid-electric buses.[22] More than 3,000 Orion buses are operating in cities across the U.S., including New York City, San Francisco, Boston, Chicago, Houston, Washington, and Seattle. The New York City Metropolitan Transportation Authority (MTA) and the Chicago Transit Authority (CTA) operate large fleets of Orion VIs, many of which were upgraded with LIBs, and Orion VII HEBs. This bus integrates a single electric motor powered by a diesel generator with the HybriDrive propulsion system.[23]

The Orion VII diesel-electric hybrid buses (Figure 2-1) with A123 LiFePO4 batteries have been in service since 2007, proving to be both energy-efficient and environmentally-friendly.[24] These batteries have greater energy density and power and are 3,000 pounds lighter than the previously-used lead acid Orion battery packs, thus improving bus fuel economy. They also have a six-year design life with lower operating and lifecycle costs. The BAE's "next generation HybriDrive" will enable more flexible modular RESS configurations with electronically-controlled cooling and APU (fuel cells or other options) to be placed on the roof for easy access and maintenance, as discussed below for fuel cell hybrid-electric buses.

BAE's HybriDrive[25] is also compatible with other hybrid and electric bus platforms and RESS options: the New Flyer Industries Xcelsior (XDE40)

---

[20]See postings at www.baesystems.com.

[21]In April 2012, DBNA ceased manufacture of Orion buses. See http://wibx950.com/daimler-buses-closing-manufacturing-operation/.

[22]BAE, "BAE Systems Hybrid propulsion Systems," presentation at AB 118 Hydrogen Workshop, California Energy Commission (CEC), September 29, 2009, www.energy.ca/2009-ALT-1/documents/2009-09-29_workshop/presentations/.

[23]See HybriDrive components and power specifications at www.hybridrive.com.

[24]See TRB/TCRP Report 132, "Assessment of Hybrid Electric Transit Bus Technology," December 2009, http://www.trb.org/Main/Blurbs/Assessment_of_HybridElectric_Transit_Bus_Technolog_162703.aspx; and "BAE/Orion Hybrid Electric Buses at New York City Transit: A Generational Comparison (NREL/TP-540-42217)."

[25]See http://www.hybridrive.com/.

40-foot diesel HEB also uses the BAE HybriDrive with LiFePO4 batteries (from Lithium Technology Corporation). The LIB delivers 200 kW peak power and is cooled with forced air. The Washington Metro Area Transit Authority (WMATA) received 152 such Xcelsior buses in 2011 and ordered 95 more in 2012 for delivery by 2013.[26]

Proterra Bus TerraVolt RESS with Altairnano Lithium Ion Titanate Battery

Proterra, Inc.,[27] has developed and fielded a broad range of lightweight composite buses that feature its TerraVolt RESS in hybrid-electric, plug-in hybrid with fast recharge, and fuel-cell-assisted battery-electric buses. Proterra hybrid, electric, and fuel cell buses use the Lithium-Ion Titanate (nLTO) battery pack from Altairnano (see Figure 2-4), consisting of stacked 50 Ah cells. The TerraVolt hybrid drive has warranty coverage for 10 years of operation, or 6,000 charge-discharge cycles. Because of its expected 18–25 years of life (longer than the typical 12–15 years bus life), this composite bus is presumed to be more cost-effective than alternative CNG buses or HEBs.

The nLTO Altairnano battery was selected by Proterra in view of its superior thermal performance, long cycle life (16,000 cycles) under deep discharge, low internal resistance that prevents overheating, and fast charging capability. The batteries can be recharged rapidly (in 5–10 minutes) at bus stops or other locations, and have a 30–40 mile range per charge. A typical battery pack consists of three 368 volt (V) parallel strings of 16 modules at 23 V, providing up to 54 kWh of useable energy (with a fourth string on reserve for increased power to 72 kWh and greater range). The large batteries are mounted in the composite floor structure to lower the bus center of gravity and confer greater stability.

The Proterra HEBs with series hybrid drive train are versatile. Proterra buses can operate in a charge-depleting (CD) battery-electric mode (BEB) for up to 20 miles, but also in deep-discharge or charge-sustaining (CS) operation if equipped with a fuel cell APU or recharged from an overhead canopy. The latter allows for range extension up to 250 miles or 20 hours of operation.

The Proterra Ecoliner BE35, deployed at Foothill Transit, is a plug-in hybrid-electric bus (PHEB) with a rooftop FastCharge hook-up to an overhead canopy or pole catenary[28] (see cover photo). Proterra has designed and manufactured its own ProDrive vehicle control and Energy Managment System (EMS). This RESS power plant requires

---

[26] See news at http://www.newflyer.com/index/2012_08_07_wmata_additional_order.

[27] See postings at www.proterraonline.com/products.asp and at www.proterra.com.

[28] See "All American Zero Emission Electric Bus Debuts on Capitol Hill," 2009, http://green.autoblog.com/2009/10/30/zero-emission-proterra-electric-bus-comes-to-capitol-hill/

active cooling, regulated by a BMS and a TMS for safety, since this RESS requires active cooling matched to the bus design and operational duty cycle.

ISE Corporation ThunderPack Integration of LIBs with Ultracapacitor APU

ISE Corporation,[29] recently acquired by Bluways, is another major developer and system integrator of series hybrid-electric drives tailored to specific bus architectures. The modular ThunderVolt electric and hybrid drives are manufactured by Siemens (ELFA) and are compatible with several hybrid-electric bus platforms,[30] including New Flyer, North America Bus Industries (NABI),[31] and Gillig hybrids.[32] The ThunderDrive RESS is flexibly-configured for diverse bus power cycling and loading requirements, integrating LIBs with APUs in fuel cell, diesel, and gasoline hybrid-electric, as well as in all-electric buses.

RESS designs are modularized for broader industry adoption and include programmable power control hardware and software and interfaces. The RESS can be optimized for diverse bus applications and routes and is placed usually on the bus roof, although it is liquid cooled. Some are LIB-based, while others include ultracapacitors (ucaps) and, if needed, blended options[33] of batteries with ucaps able to rapidly store and deliver on-demand energy from either an APU or from saved regenerative braking energy.

Since 2002, ISE has entered into a strategic development and supply agreement with Maxwell Technologies, a San Diego manufacturer of the PowerCache ucaps.[34] The ThunderPack (ucap) RESS uses 150 large-cell ucaps to rapidly store and release 150 kW, or as a dual unit up to 300 kW of power for hundreds of thousands of cycles.

Since 2007, ISE has integrated an LIB in RESS, either LiFePO4 from A123 or LTO from Altairnano. The ThunderPack RESS for battery-dominant hybrid buses integrates LIBs with a TMS located on the roof of the bus and with compact, scalable, high-energy Maxwell BoostCap ucap modules (Ultra-E 500). This combination RESS enables up to a million charge/discharge cycles for rapid capture of energy (regenerative braking) at high power levels. To date, the ThunderVolt with LIBs has been deployed in more than 500 in-service buses and has operated for more than 10 million miles cumulative, including the following:[35]

- Diesel-hybrid articulated 62-foot StreetCar rapid transit buses built by Wright in Las Vegas.

---

[29] See postings at http://www.isecorp.com/hybrid-technologies/.

[30] See hybrid-electric drive subsystems posted at www.isecorp.com/applications/transit-bus.

[31] See http://www.nabusind.com/index.asp.

[32] See http://www.gillig.com/New%20GILLIG%20WEB/hybrid.htm.

[33] See http://www.isecorp.com/energy-storage/lithium-ion-power-systems/.

[34] See the ISE and Maxwell Integrated Energy Storage systems at www.isecorp.com/energy-storage/.

[35] Paul B. Scott, ISE Corp., presentation at CEC Hydrogen Workshop, Sept 29, 2009, www.energy.ca.gov/2009-ALT-1/documents/2009-09-29_workshop/presentations.

- More than 250 New Flyer gasoline-hybrid buses, operated by more than 10 U.S. transit operators, which have accumulated more than 12 million in-service miles. These buses have been operated by the San Diego Metropolitan Transit System (MTS) since 2008. Their RESS integrates batteries (option includes LIBs for battery-dominant hybrids) with electric TMS are located on the bus roof with compact, scalable, high-energy ucap modules (Ultra-E 500).
- ISE/New Flyer Hybrid Hydrogen Internal Combustion Engine (HHICE), introduced in 2004 by SunLine, combined the ISE ThunderVolt integrated electric propulsion system using Altairnano nLTO LIBs with a hydrogen-fueled ICE.
- Fuel cell hybrids—in 2004–06, Alameda Contra Costa Transit (AC Transit) and SunLine Transit (serving the Palm Springs, California, area) demonstrated 40-foot Van Hool buses using UTC Power's fuel cell stacks and compressed hydrogen fuel integrated with an ISE hybrid-electric drive system.
- The ongoing American Fuel Cell Bus (AFCB) project (discussed in Section 3) will demonstrate an improved ISE electric traction and power electronics system, featuring UTC fuel cells (120 kW) and the compact high-power EnerDel LIB, in a lighter and quieter New Flyer 40-foot bus. This FCB architecture was deployed in 20 New Flyer fuel cell hybrid buses for the Vancouver 2010 Winter Olympics, but the 5 Wright FCBs in London were grounded during the 2012 Summer Olympics for security reasons.

Figure 3-1 shows how the ISE hybrid drive is placed in the complex architecture of a fuel cell hybrid bus. Figure 3-2 shows the modular LIBs linked to the ISE BLUWAYS proprietary BMS/TMS controller module.[36]

Figure 3-1

BLUWAYS/ISE Hybrid Drive integrated in fuel cell hybrid bus

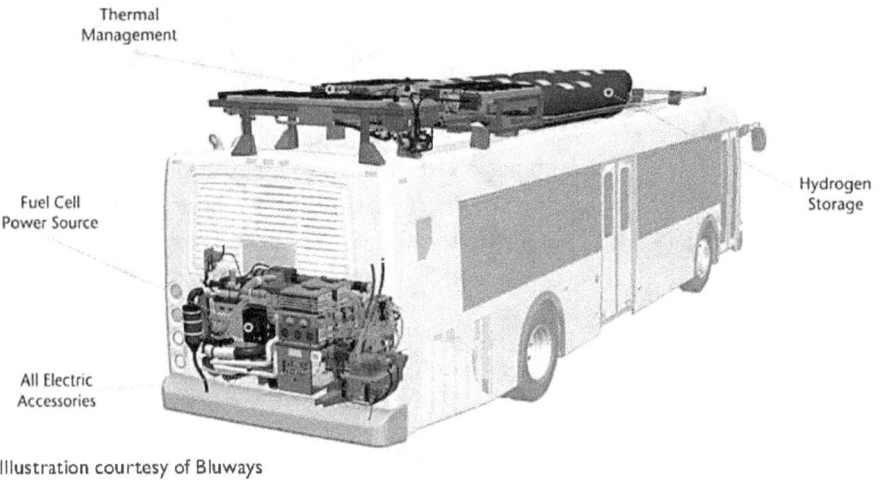

Illustration courtesy of Bluways

---

[36]See details of hybrid drive and RESS architectures posted at www.isecorp.com/hybrid-technologies/and www.isecorp.com/energy-storage.

SECTION 3: EXAMPLES OF HYBRID AND BATTERY ELECTRIC TRANSIT BUSES WITH LIB-BASED RESS

Figure 3-2
LIB modules in BLUWAYS ISE energy storage system, with master control module attached

Illustration courtesy of Bluways

DesignLine EcoSaver IV HEB with LIB and Capstone Microturbine APU

The DesignLine EcoSaver IV bus was specifically developed for U.S. markets.[37] It is an HEB with a lightweight aluminum body with twice the fuel efficiency, but half the tailpipe emissions relative to a conventional diesel bus. The EcoSaver bus was successfully demonstrated in 2007 by the Chicago Transit Authority and was evaluated in New York City in 2009 where it did not meet route requirements. In 2009, the City of Baltimore purchased 21 EcoSaver buses for use as Charm City circulators after a successful demonstration. This HEB uses a Capstone (30 kW) micro-turbine diesel engine and generator as an APU. The Capstone micro-turbine is usually mounted in the back of the bus, while the Altairnano nLTO batteries are placed in the bus floor behind and between the wheels. The micro-turbine APU is used as a range extender. It produces alternating current power to recharge onboard LIBs via two inverter (250 kWs each). The LIB, in turn, powers two 120 kW, three-phase, AC induction motors for traction. The EcoSaver bus operation and RESS are managed by an advanced BMS with APU controls and driver interface (see Figure 3-3).

---

[37] See bus specifications at http://www.designlinecorporation.com/EcoSaver%20IV.%20pdf.pdf.

Figure 3-3

Charm City Circulator, a DesignLine EcoSaver IV operating in downtown Baltimore

Photo courtesy of DesignLine Corporation

Enova Systems HybridPower for School and City PHEBs

Enova Systems[38] has developed and markets a suite of HybridPower drives and related power conversion and management components for series and parallel hybrid and all-electric buses. Enova's HybridPower drive train was integrated by IC Corporation[39] (bus division of Navistar, Inc.) in both CD and CS plug-in hybrid post-transmission parallel diesel-electric buses. In the Enova CD hybrid buses, there are several RESS battery options for the CD PHEB school bus: NiMH, PbA, or dual LIB packs. The latter is integrated within the hybrid cooling package and must be recharged overnight. These battery packs can be deep-discharged to about 25 percent State of Charge (SOC) over the 40-mile range after overnight recharging. In the Enova CS hybrid buses, the CS RESS system for the latest ZEB school bus integrates the Enova post-transmission 80 kW hybrid drive train with a LIB pack and an electric motor. The battery SOC is maintained by onboard equipment and does not require grid recharge interfaces.

Since 2007, Enova/IC post-transmission parallel PHEBs have operated in California and Florida school districts. In 2009, DOE funded Enova and IC/Navistar with federal stimulus dollars to deploy 16 PHEB school buses nationwide. It couples a diesel engine with an 80 kW AC induction electric motor mounted behind the transmission (post-transmission) and a hybrid controller with a battery care unit (BCU) to monitor the battery voltage, SOC, amp-hours,

---

[38]See Enova bus specifications and ESS at www.enovasystems.com and http://www.enovasystems.com/drive-system-components.html.

[39]See www.ic-corp.com postings and http://eon.businesswire.com/news/eon/20100526005345/en.

kWh, and temperature. The BCU controls a Safety Disconnect Unit (SDU) to ensure the safety of the battery pack during charging and provides surge protection and automatic disconnect in case a ground fault occurs.

Enova Systems also developed a battery-dominant FC power train for HEBs and PHEBs. This drive system integrates an induction electric motor, a dual 8 kW inverter, a 380 V DC/DC converter, a BCU with safety disconnect protection, and a digital PMS. This RESS can supply modular power (90, 120, or 240 kW) using an EMS. The Enova hybrid drive provides 120 kW of power using as APU a 20 kW Hydrogenics FC stack. Enova battery-dominant FC shuttle buses[40] were demonstrated at Hickam Air Force Base in Honolulu, Hawaii, and evaluated by NREL. Such FC HEBs currently operate in the Denali National Park in Alaska using the Valence Technology XP-U-Charge Saphion® Lithium Ion Phosphate Batteries for improved safety and thermal stability.

Enova's pre-transmission hybrid drive systems were also delivered to First Auto Works, a large Chinese bus and vehicles manufacturer, for their Liefang 12-meter buses (103 passenger, 85 kilometers per hour top speed), due to deploy and operate in 13 Chinese cities.

FTA Advanced Transit Bus Demonstration and Deployment Programs

National Fuel Cell Bus Program (NFCBP)

FTA's NFCBP is a cooperative research, development, and demonstration program started in 2006 to advance the technology and enable the viability of a commercial FCB. Over the past decade, the NFCBP has funded the development, nationwide demonstration, and commercialization of advanced FCBs with diverse architectures, in cooperation with nonprofit consortia including DOE, CALSTART, Inc., Northeast Advanced Vehicle Consortium (NAVC), and state and local agencies. These zero-emission FCBs have steadily improved by combining hydrogen fuel cells for auxiliary power (with hydrogen currently produced by solar electrolysis in California) with hybrid-electric drive, now storing energy in advanced LIBs. NREL and FTA have evaluated their transit operational performance, fuel economy, cost, reliability, and evolutionary technology improvements.[41] Second-generation FCBs now under development include FC-dominant CS designs (FC recharges smaller batteries), battery-dominant CD options (FC acts as an APU range extender), and diesel hybrids with a small FC

---

[41]See Hydrogen Fuel Cell Bus Evaluations and Plans, http://www.nrel.gov/hydrogen/proj_fc_bus_eval.html and http://www.nrel.gov/hydrogen/pdfs/49342-2.pdf.

powering accessories.[42] NREL has posted evaluations and project fact sheets that provide details on the RESS architecture and specifics of LIB chemistry and suppliers for FC hybrid buses.

As of mid-2011, 25 FC transit buses were in service in the U. S, including:
- 18 Van Hool buses with UTC Power's fuel cells
- 1 New Flyer bus with a Ballard fuel cell
- 2 Proterra plug-in hybrids with Hydrogenics fuel cells
- 3 Ebus plug-in hybrids with Ballard fuel cells
- 1 Daimler/BAE diesel hybrid with a Hydrogenics fuel cell APU

In April 2012, U.S. DOT and FTA awarded $13.1 million for 11 new FCB projects nationwide, of which 5 were to CALSTART, Inc.[43] for $6.6 million to demonstrate California ZEB advanced concepts with more powerful and smaller fuel cells and more efficient power electronics and components.

Seven additional buses will be added to the U.S. transit bus demonstration fleet by FTA's NFCBP. These new FCBs will use hydrogen fuel cells in combination with advanced lithium-ion batteries for RESS energy storage and regenerative braking, some in a battery-dominant configuration with the FC as an APU.

The AFCB project[44] will add a more advanced FCB to the SunLine (Thousand Palms, California) transit fleet. Its partners are FTA, CALSTART, BAE Systems as the series hybrid power and propulsion system integrator, El Dorado National and Axess as the bus manufacturers, and Ballard Power Systems 150 kW as the fuel cell manufacturer. In 2011, SunLine also received an FTA TIGGER grant award for two additional FCBs. El Dorado National will manufacture and deliver the buses by 2013.[45]

Figure 3-4
SunLine Fuel Cell Bus

Photo courtesy of L. Eudy, NREL

---

[42] See U.S. Fuel Cell Bus Deployment summary, www.fuelcells.org/info/charts/fcbuses-US.pdf.

[43] See "FTA Announces 5 Awards to CALSTART, Inc., Advancing Zero Emission Bus Technology," April 10, 2012, www.calstart.org/News_and_Publications/.

[44] See Fact Sheet at www.nrel.gov/hydrogen/pdfs/nfcbp_fs4_feb12.pdf.

[45] See http://www.sunline.org/sunline-receives-fta-award-for-two-additional-fuel-cell-buses.

SECTION 3: EXAMPLES OF HYBRID AND BATTERY ELECTRIC TRANSIT BUSES WITH LIB-BASED RESS

The Bay Area consortium led by AC Transit will operate a fleet of 12 Van Hool FCBs with ISE ThunderVolt hybrid drive and UTC Power 120 kW fuel cells. Although the earlier (2006) AC Transit HyRoad FCBs used three ZEBRA (Nickel Sodium Chloride) batteries, the latest use the Siemens ELFA e-Drive hybrid propulsion with EnerDel LIBs. The EnerDel LIB chemistry features a Lithium Manganese Oxide (LMO) cathode and a Lithium Titanate (LTO) anode with a non-flammable electrolyte to mitigate potential thermal runaway and fire. These heavy-duty traction batteries (rated capacity 29 ampere-hours and rated energy 17.4 kWh) are packaged in compact flat foil-wrapped pouches with a large surface area that facilitates cooling.[46] The EnerDel LIBs are integrated in a similar RESS configuration for the NFCBP-funded Nutmeg Project Fuel Cell Bus,[47] to be demonstrated in Connecticut by CT Transit.

The San Francisco Compound Bus 2010 FCB is part of the Bus 2010 FTA and CALSTART, Inc., initiative[48] and aims to double the fuel efficiency of a diesel bus in an affordable package. It links a fuel cell, conventional diesel engine, and battery energy sources into one power and propulsion system. It combines the Daimler/BAE Orion VII diesel hybrid-electric propulsion with a LIB (200 kW peak power), complemented by two Hydrogenics FCs (12 kW each) fueled by four hydrogen tanks that boost the small diesel engine bus power and extend its range.

The NFCBP also funded Proterra in 2007 to design and integrate its TerraVolt RESS into a 35-foot lightweight composite body for the Hydrogen Fuel Cell 35-foot (HFC35) battery-dominant FC HEB. FTA also funded demonstrations of this HFC35 lightweight bus in Columbia, South Carolina. The RESS supplies electric power to two UQM Technologies PowerPhase 150 kW electric propulsion motors. This electric motor[49] is a brushless permanent magnet (PM) motor/generator able to recover up to 90 percent of the braking kinetic energy and convert it to electrical energy. This configuration allows for flexible bus operation in either a battery-powered electric mode, or using an FC APU for range extension.

In a battery-dominant FCB, the LIB is recharged by an APU consisting of two small, 16 kW Hydrogenics polymer electrolyte membrane (PEM) fuel cells (fueled by pressurized hydrogen in tanks on the bus roof). The Proterra battery-dominant FCB has operated for the past few years in Burbank, California, logging about 100 miles per day with the original RESS, thus proving its reliability and durability.

---

[46] See NREL report, "Fuel Cell Buses in U.S. Transit Fleets: Current Status 2010—AC Transit," www.actransit.org/wp-content/uploads/NREL_rept_OCT2010.pdf.

[47] See Nutmeg CT Transit Fuel Bus project NFCBP factsheet at http://www.nrel.gov/hydrogen/pdfs/nfcbp_fs3_jul11.pdf.

[48] See "San Francisco Hosts National Fuel Cell Bus Program Demonstration" at http://www.nrel.gov/hydrogen/pdfs/nfcbp_fs2_jul11.pdf.

[49] For information on UQM Technologies electric motors for hybrid and electric buses, see http://www.uqm.com/pdfs/PowerPhase150%20_edited_.pdf.

SECTION 3: EXAMPLES OF HYBRID AND BATTERY ELECTRIC TRANSIT BUSES WITH LIB-BASED RESS

SunLine Transit Agency[50] (in Thousand Palms, California) has demonstrated in service several FCB architectures using lessons learned to improve succeeding generations. The latest SunLine Advanced Technology (AT) FCB from New Flyer uses the Bluways USA[51] hybrid-electric drive with Siemens ELFA components. It stores regenerated braking energy in Valence Technology Lithium Iron Phosphate batteries (47 kWh) and uses auxiliary power from Ballard Fuel Cells (HD6, 150 kW) fueled by 6 tanks of 5,000 psi hydrogen (see Figure 3-4).

The battery-dominant Lightweight Hybrid Fuel Cell Bus (LWFCB) is an innovative NFCBP bus prototype under development for demonstration by General Electric (GE).[52] Its dual battery RESS consists of two complementary batteries: an LIB with high power but low energy-storage capacity (a "power battery"); and two 18 Wh, 278 V sodium metal halide (NaMX) Durathon batteries with high energy density but lower power. This dual battery will be able to meet the required 60 kWh energy storage capacity and 160 kW peak power load at lower cost than a scaled-up, large, and expensive LIB.

FTA TIGGER and Clean Fuels Grant Programs

FTA Research, Development and Technology (RD&T) and synergistic Clean Fuels grant programs have enabled the development, demonstration, evaluation, and commercialization of hybrid and electric drive transit buses that integrated heavy-duty, high-performance LIBs in diverse power-train architectures. Two notable programs are the three-year (2009–2011) TIGGER[53] and Clean Bus/Clean Fuels grants augmenting the Bus and Bus Facilities awards.[54] These FTA competitive grant awards to public transit agencies will lead to substantial expansion of hybrid and electric transit bus fleets and will permit demonstration and evaluation of novel RESS systems for in-service operation over a wide range of climates and routes.

The TIGGER program's goal is to promote the deployment of innovative technologies that improve the energy efficiency and reduce the environmental emissions of transit facilities and fleets. Three annual cycles of proposal evaluation and awards have funded numerous innovative transit bus projects nationwide. TIGGER program awards will advance the state of the art for electric drive bus technology and improve safety and performance at lower cost in the future.

---

[50]See L. Eudy and K. Chandler, NREL, "SunLine Transit Agency Advanced Technology Fuel Cell Bus Evaluation: Second Results Report and Appendices," October 2011, http://www.nrel.gov/hydrogen/pdfs/52349-2.pdf.

[51]Bluways USA is the current owner of ISE Corporation technologies. See http://www.isecorp.com/company/ise-timeline/.

[52]See L. Salasoo, T. Richter, R. King, and Z. Li, SAE Technical Paper 2012-01-1029, "GE Electric Drivetrain Technologies for Lightweight Battery Dominant FCB," at http://papers.sae.org/2012-01-1029/.

[53]See postings and awards for the three-year TIGGER program at http://www.fta.dot.gov/%20TIGGER.

[54]See list of FTA grant programs for clean bus technologies, fuels and facilities at http://www.fta.dot.gov/grants/13094.html.

SECTION 3: EXAMPLES OF HYBRID AND BATTERY ELECTRIC TRANSIT BUSES WITH LIB-BASED RESS

NREL is currently conducting TIGGER project evaluations for FTA to shed light on the actual in-service performance attributes of these upgraded bus fleets. Posted project information sheets[55] illustrate the wide range of hybrid, electric, and fuel cell bus designs, manufacturers, technologies, drive architectures, and charging infrastructure characteristics of buses with LIBs to operate in diverse U.S. locations, routes, and climatic conditions. Selected examples of recent TIGGER I awards to deploy HEB/EB/FCBs integrating LIBs include the following:

- Along with TIGGER II funding, Foothill Transit of Pomona, California, will acquire nine new electric Proterra buses and expand recharging infrastructure to complement the three buses it has successfully operated for the past three years.
- VIA Metropolitan Transit in San Antonio, Texas, will acquire three Proterra all-electric battery buses with fast-charge infrastructure powered by wind and solar renewable generation.
- Metro Mobility of Minneapolis-St. Paul acquired smaller Azure Dynamics (AZD) gasoline-electric parallel hybrid buses with the AZD Balance Hybrid Drive and Forcedrive LIBs with nickel cobalt aluminum (NCO) chemistry from Johnson Controls-Saft.
- Community Transit in Everett, Washington, will acquire 15 New Flyer diesel-hybrid buses with the BAE Systems HybriDrive and powerful LIBs that maximize energy storage and reduce engine size.
- Link Transit in Wenatchee, Washington, will acquire up to 10 electric trolley buses manufactured by Ebus with Altairnano nLTO batteries as well as 2 quick-charge stations powered by renewable hydropower.

---

[55] See project details listed at www.fta.dot.gov/about_FTA_14440.html.

SECTION 4

Lessons Learned, Progress, and Prospects Overview of Transit Buses with LIBs

Case Studies and Safety Lessons Learned from LIB Bus Operations

Valuable lessons learned regarding LIB battery capabilities, reliability, and safety were gained from transit agency operational experience with hybrid and electric drive transit buses. Although some battery maintenance and reliability problems encountered were typical new technology "growing pains," others posed safety hazards. The DOE's Alternative Fuels Data Center (AFDC) database for hybrid and electric heavy-duty vehicles and engines[56] contains information on more than 15 diesel and gasoline electric hybrid and electric buses that are commercially available from multiple manufacturers and suppliers. Transit buses with a wide range of integrated hybrid propulsion systems, most of which feature LIB-based RESS, are now available new or repowered. Models in U.S. operations include the Daimler Bus NA Orion VII, DesignLine EcoSaver IV, El Dorado National Axess, Foton America FCB, Gillig (with Allison dual-mode compound split HEB), NABI diesel electric with ISE ThunderPower, Proterra electric or fuel cell buses with TerraVolt and Altair Lithium Ion Titanate (LTO) LIBs, and the New Flyer Xcelsior hybrids with BAE or ISE LIBs. New choices are emerging, such as the Chinese North America Build Your Dream (BYD) electric eBus that will serve as shuttles at the Los Angeles airport.[57]

Information on LIB battery failures and maintenance issues can also be found in NREL evaluations of hybrid and electric transit bus fleets operating in many U.S. cities.[58,59] These reports evaluate hybrid and electric bus fleets such as the New York City MTA's BAE Orion diesel-hybrid bus fleet, Long Beach Transit's gasoline-electric hybrids, King County Metro's Allison hybrid-electric buses, and Knoxville Area Transit's Ebus electric buses and trolleys. For instance, the NREL

---

[56]See detailed searches at www.afdc.energy.gov/afdc/vehicles/search/heavy/engines.

[57]See press release at http://www.engadget.com/2011/10/25/byd-opens-north-american-hq-in-la-electric-bus-headed-for-lax/.

[58]See transit hybrid fleet reports posted at http://www.afdc.energy.gov/afdc/fleets/transit_experiences.html?print.

[59]See postings listed at www.afdc.energy.gov/afdc/fleets/transit_experiences.html.

SECTION 4: LESSONS LEARNED, PROGRESS, AND PROSPECTS OVERVIEW OF TRANSIT BUSES WITH LIBS

multi-generational comparison of Orion/BAE hybrid-electric buses operating in the New York City MTA's transit fleet[60] identified the key LIB performance, durability, and safety improvements needed for transit reliability, availability and durability.[61]

The safe operability of transit buses with RESS based on LIBs is a key concern being addressed by U.S. DOT regulatory and oversight programs within FTA, the National Highway Traffic Administration (NHTSA), and the Federal Motor Carrier Safety Administration (FMCSA), as well as by state agencies. LIBs may pose potential risks due to thermal runaway when overheated, or even lead to explosion and fire when the cell is ruptured.

Flammable lithium and vented hydrogen or oxygen may promote fires, and corrosive or toxic electrolyte could leak if the battery is breached. These hazards are preventable through design, packaging, and abuse testing, and by using voltage and temperature monitoring systems for shutdown and electrical isolation. BMS, TMS, mechanically-crashworthy packaging, and active or passive cooling of the RESS subsystem assist in alleviating these risks.

The electrical safety of hybrid and electric vehicle high-voltage batteries in general, and the prevention and mitigation of fire and explosion hazards associated with LIBs in particular, are being addressed by ongoing NHTSA research and regulatory programs. The May 2012 NHTSA Electric Vehicle Safety Symposium discussed the LIB system integration and operational safety assurance strategies among other potential hazards emerging for commercial hybrid, plug-in hybrid, and electric vehicles.[62]

Standards Developing Organizations (SDOs) with specific focus on LIB hazards and safeguards include SAE, NFPA, and Underwriters Laboratory (UL).[63]

Two Electric Vehicle Safety Summits have been held to date[64] by stakeholders, including industry, NHTSA, NFPA,[65] and SDOs (ANSI, SAE, UL), and have identified the research gaps and training needs for emergency responders. The NHTSA safety LIB analysis research program[66] identified safety-critical failures. Prevention and mitigation of hazards such as thermal runaway events can be

---

[61] See Clean Air Initiative: Infopool–Hybrid Bus postings at www.cleanairnet.org/infopool/1411/propertyvalue-17735.html#h2_5.

[62] See NHTSA overview and presentations posted at www.nhtsa.dot.gov/vehicle+Safety/Electric+Vehicle+Safety+Symposium.

[63] See FTA Transit Safety information at http://bussafety.fta.dot.gov/splash.php and Volpe Center transit safety resources posted at http://www.transit-safety.volpe.dot.gov/Safety/Default.aspx.

[64] See postings at http://www.evsafetytraining.org/News/News-Articles/NFPA-SAE-Summit.aspx.

[65] See Fire Protection Research Foundation, "Lithium Ion Batteries Hazard Assessment," NFPA, July 2011, at www.nfpa.org/Foundation.

[66] See 2012 presentations at http://www.unece.org/fileadmin/DAM/trans/main/wp29/WP29-155-43e.pdf and NHTSA-Battelle report overview at http://www.sae.org/events/gim/presentations/2012/stephensbattelle.pdf.

achieved through thermal management and packaging and improved electrical isolation and crashworthiness of battery modules and subsystems. Existing regulations, such as the Federal Motor Vehicle Safety Standard No. 305 (Electric powered vehicles; Electrolyte spillage and electrical shock protection) battery subsystem safety regulation, and numerous existing and emerging SAE standards[67] will improve and ensure battery safety in hybrid and electric vehicles. Safety training programs of maintenance workers, first responders, and fire marshals were also developed and implemented to complement Materials Safety Data Sheets (MSDS) for RESS provided by Original Equipment Manufacturers (OEM) or battery suppliers.

On October 28, 2011, NHTSA's Office of Defect Investigation (ODI) issued a safety recall and corrective modifications or replacement of LIBs in Daimler Orion VII hybrid-electric buses.[68] After several incident investigations due to potential breaching of electrical isolation hazards from the accumulation of debris and moisture,[69] 1,300 Orion VII hybrid buses with the BAE HybriDrive and RESS using A123 LiFePO4 batteries (manufactured in November 2008), and some earlier 2006–07 models retrofitted with LIBs, were recalled to replace battery modules on the bus roof.

The BAE Systems HybriDrive[70] chose a LiFePO4 battery chemistry (developed by A123 Systems) for its superior thermal stability and proven operational safety, as well as for its modular, compact, lightweight design, and long cycle life (over six years). The roof placement did not require active battery cooling, but roof placement also allowed the debris and moisture accumulation to cause potential short-outs and fires.

This recall affected more than 1,600 Orion buses operating in New York City. The 2011 Orion VII safety recall has already affected many urban transit fleets in the U.S. and led to Daimler's discontinuing the manufacture and marketing of Orion VII hybrid buses in the U.S. and Canada in April 2012.[71]

The New Flyer manufacturer of Xcelsior buses, which also use the BAE Hybridrive RESS with A123 LIBs, notified NHTSA in March 2012 that it was recalling them to correct a similar RESS that could cause a LIB short and pose fire

---

[67] See SAE/GIM presentation, January 2012, at http://www.sae.org/events/gim/presentations/2012/galyenmagna.pdf on J1766, "Recommended practice for electric and hybrid vehicle battery systems crash integrity testing."

[68] See October 28, 2011, http://recallcast.com/recalls/2011/oct/28/daimler-buses-noth-america-inc-electr-11v523000/.

[69] See Orion 7 Hybrid Bus Recall at http://recallcast.com/recalls/2011/oct/28/daimler-buses-north-america-inc-electr-11v523000/ and at http://www.safercar.gov.

[70] See www.hybridrive.com/hybrid-transit-bus.asp and www.hybridrive.com/lithium-ion-energy-storage-system.asp.

[71] See http://www.newschannel6now.com/story/17736988/daimler-buses-reconfigures-operations-in-north-america?clienttype=printable.

hazards. A total of 47 hybrid Xcelsior Metrobuses in the Washington, D.C., area were also pulled from operation for inspections and corrective retrofits of LIBs by BAE Systems.[72]

Another LIB-related case study impacts the DesignLine EcoSaver IV hybrid-electric buses, which feature GAIA LIBs and Capstone micro-turbine APUs and also experienced mechanical and reliability problems. A total of 21 DesignLine Charm City Circulator[73] buses in Baltimore were removed from service. There were some DesignLine EcoSaver IV bus order cancellations by the Denver Regional Transportation District and by the New York City MTA after their in-service testing of five HEBs revealed that the EcoSaver buses were underpowered and unreliable for expected routes and duty cycles.

The most important lesson learned from these case studies is that LIB reliability and safety are of paramount importance for hybrid and electric buses using LIBs for energy storage in their hybrid drive for electric propulsion.

To date, both the probability and severity of incidents involving LIBs overheating, degassing, or having electrical short-out incidents were very low. Only 10 similar incidents have occurred in the U.S. and Canada in the past decade out of more than 2,200 operating hybrid buses, with no resulting injuries. This "learning curve" is typical for the technology shake-down period in field operations and will lead to further design improvements for enhanced safety, reliability, and maintainability of advanced RESS systems in transit bus fleets.

LIB Bus Market Prospects and Challenges

It is clear that the U.S. and global hybrid and electric transit bus industry is ready to use improved LIBs if and when the emerging batteries become proven in performance, durability and cost, and commercially viable. Thanks to the recent progress and federal investments in fuel-efficient transit buses, there are diverse choices of LIBs and RESS integration available to transit authorities able to renew their bus fleet with cleaner options. NHTSA bus safety regulatory oversight, as well as early FTA and transit agency acceptance testing programs, can and will avert LIB-related hazards and correct faulty RESS designs or LIB products, enabling continued industry growth and the adoption of innovative technologies in bus applications.

Transit buses with a wide range of integrated hybrid propulsion systems, most of which currently feature LIB options for superior traction energy and power,

---

[72] See news item at http://washingtonexaminer.com/local/transportation/2012/03/47-hybrid-metrobuses-recalled-flawed-battery/416416.

[73] "For Charm City Circulator, Growing Pains are Inevitable," www.bizjournals.com/baltimore/stories/2010/08/23/.

are now available as either new, or repowered for diverse sizes and traction power.

Challenges to large-scale LIB integration into existing bus fleets remain, although progress in extending LIB cycle life, reducing costs, improving performance reliability, and ensuring safety in operations and maintenance is rapid. The LIB operational challenges that need to be fully addressed to enable further commercialization and transit fleet adoption include:

- Monitoring and control of LIB cells voltage balancing and conditioning (equalization in the stack).
- Extending the cycle life of LIBs, which now is warrantied for less than half of bus life.
- Redesign of hybrid-electric drive components that failed early.
- Stable control systems for traction-power electronics (hardware and software).

As illustrated, there has been rapid and steady progress[74] in ensuring the performance reliability and safe operability for a diverse array of vehicular LIBs integrated in a wide range of propulsion and RESS architectures, routes, duty cycle requirements, and climate conditions.

Battery-specific safety training for maintenance staff and bus operators is also needed to complement the automated RESS monitoring and control systems.

Another challenge is how to extend the rather short lifecycle of large-format LIBs compared to previously-used heavier and larger, but cheaper NiMH, PbA, or ZEBRA batteries, especially for CD operating cycles. The best LIB design life is only 6 years (under warranty coverage), although the average design and operating life for a transit bus is 12–15 years. In practice, battery failures sometimes occurred sooner (1–2 years).

The high cost of LIB cells, modules, and stack packaging is another major challenge; a battery pack replacement could cost up to $60,000. Greater competition will continue to lower LIB costs, and LIB chemistries, packaging, and quality control, coupled to better monitoring systems, promise to resolve LIB reliability and longevity issues. In this globally-competitive LIB market, long-term maintenance contracts, including free LIB replacement and longer-duration factory warranties, are needed to offset the high initial cost and replacement cost

---

[74]See "Live Reporting from Advanced Automotive Battery Conference (AABC) 2012: Latest Battery Technology Developments" at http://www.cars21.com/content/articles/75120120210.php and "AABC2012: Challenges and Solutions for Cost-Effective Integration of Batteries into Electric Vehicles" at www.cars21.com/content/articles/75220120213.php.

for LIBS. A good strategy to ensure safety for the LIB subsystem is to negotiate with battery manufacturers and suppliers for long-term warranties (3–6 years) instead of the typical 1–3 years.

Continuing progress is also needed in ruggedizing the battery pack and LIB abuse tolerance to ensure the containment of electrolyte, or outgassing due to thermal stress in extreme weather, or from mechanical failure in road crashes.[75] Integrating the electronic sensors for LIB monitoring with fault detection and diagnostics software, managing TMS and BMS performance, and detecting internal aging and degradation over time, will help assure the LIB structural stability. These challenges will probably be overcome in synergy with the growing market for light-duty electric, hybrid, and plug-in hybrid vehicles.

There is great promise: The DOE Battery R&D program, in cooperation with battery makers, has targeted improvements in next-generation LIB materials and coating for better abuse tolerance and cell durability. They include higher-capacity and new materials for anodes (like silicon and metal alloys), higher voltage and capacity cathodes, non-flammable electrolytes, resilient separators, and sturdy packaging. Prospects for the widespread commercialization and integration of advanced high-power and high-energy LIBs in bus fleets are improving as a result of federal research investments and subsidies for manufacturing of automotive LIBs. Increasing competition among battery suppliers for automotive applications will lower the cost per kWh and drive performance enhancements.

The 2012 DOE Annual Performance Review presentations[76] on Energy Storage R&D highlighted continuing progress and the ambitious technical targets for battery development and demonstration. DOE has awarded contracts to multiple battery industry leaders (A123 Systems, LG Chem., Saft-Johnson Controls) and to the U.S. Advanced Battery Consortium (USABC). A next generation of LIB's is under active research and development and includes lithium-sulfur, lithium-air, and lithium-metal formulations with higher-capacity anodes and higher-voltage cathodes.

NHTSA research[77] is underway to assess LIB safety issues during and after vehicle crashes and the prevention and mitigation of battery-related fires and electrocution hazards. Guidance is emerging from standards organizations developing battery testing, certification, and recharge interfaces and by the battery and automotive manufacturers that provide warranties.

---

[75] See Exponent, "Lithium Ion Batteries Hazard and Use Assessment," final report for NFPA, 2011, at http://www.nfpa.org/assets/files/pdf/research/rflithiumionbatterieshazard.pdf.

[76] See David Howell, review presentation of battery R&D activities by DOE/EERE Vehicle Technologies Program (VTP), May 14, 2012, at http://www1.eere.energy.gov/vehiclesandfuels/pdfs/merit_review_2012/plenary/vtpn07_es_howell_2012_o.pdf.

[77] See NHTSA Electric Vehicle Symposium and Interim Guidance documents, May 18, 2012, posted at http://www.nhtsa.gov/Vehicle+Safety/Electric+Vehicle+Safety+Symposium, and "Failure Modes & Effects Criticality Analysis of Lithium-Ion Battery," SAE/GIM presentation at www.sae.org/events/gim/presentations/2012/stephensbattelle.pdf.

The hybrid and electric bus and other heavy-duty vehicles industry is currently in transformation in order to comply with the new and stricter NHTSA and EPA heavy-duty vehicles fuel efficiency and greenhouse gas emission standards for 2014–2018 and proposed for 2018–2025.[78] Next steps that are already underway will enable larger-scale transit bus fleet hybridization and electrification, as well as improved fuel efficiency and environmental sustainability such as the following:

- Faster recharging of RESS in station or while moving "on the fly"—the Proterra EcoLiner BE35 buses use Fast Fill recharge of Lithium Titanate batteries in 5–10 minutes at station stops, affording ranges of 30–40 miles of operation per single charge. The next step in speed and convenience will be even faster and allow wireless charging of buses while moving, either by induction or magnetic resonance from coils embedded in the roadbed, or from overhead structures. FTA's TIGGER program has selected several innovative wireless power transfer projects, such as the Chattanooga Area Transit Authority (CARTA), which will develop, demonstrate, and evaluate rapid inductive charging of three electric buses from roadbed coils with a potential 95 percent efficiency.[79] Another innovative TIGGER project involving wireless recharging of an electric bus by magnetic resonance was awarded to the University of Utah using WAVE, Inc., technology.[80] Shaped magnetic resonance charging (such as that proposed by WiTriCity and OLEV On Line Electric Vehicle) is another wireless charging technology option. A TIGGER III award to the City of McAllen, Texas, transit[81] will introduce and evaluate the OLEV shaped-magnetic-field-in-resonance (SMFIR) wireless charging technology[82] to convert three diesel buses to all-electric, rapidly-recharged, energy-efficient, and range-extended operation. The advantage is that the RESS needed is three times smaller, lighter, and also cheaper than for existing electric buses.

- A valuable next step for electrification of transit buses is to develop wireless charging standards. SAE is now developing, in partnership with several technology providers, Standard J2954 for wireless charging of plug-in hybrids and electric vehicles.[83] Inductive power charging of hybrid and electric buses is already operational in Japanese and Italian cities.[84]

---

[78] See NHTSA and EPA heavy-duty vehicle fuel economy regulations posted at www.nhtsa.gov/fuel-economy/.

[79] See "CARTA Electric Buses to Charge on the Go in Chattanooga," November 18, 2011, at http://www.timesfreepress.com/news/2011/nov/18/cartas-electric-buses-charge-go/.

[80] See "Electric Bus Charges Wirelessly at U of U," November 16, 2011, at http://www.ksl.com/?nid=148&sid=18116082&title=electric-bus-charges-wirelessly-at-u-of-u.

[81] See "McAllen, TX, to Introduce OLEV-powered Electric Buses" at http://evworld.com/news.cfm?newsid=27074.

[82] See postings at wwwolevtech.com and http://www.witricity.com/pages/technology.html.

[83] See presentation by Jesse Schneider, SAE TIR J2954 Chair, on "Wireless Charging of Electric and Plug-in Hybrid Vehicles" at http://bioage.typepad.com/files/SAE%20J2954%20Wireless%20Charging%20Dec.%202010.pdf Industry; includes WiTriCity in the U.S., Conductix Wampfler in Germany, HaloIPT in the UK and New Zealand.

[84] See "In Italy electric buses pick up wirelessly their power," *The New York Times*, May 30, 2012, at http://wheels.blogs.nytimes.com/2012/05/30/in-italy-electric-buses-wirelessly-pick-up-their-power/.

SECTION 4: LESSONS LEARNED, PROGRESS, AND PROSPECTS OVERVIEW OF TRANSIT BUSES WITH LIBS

- Next steps are also underway to develop LIBs with higher capacity and power. High performance RESS using ultracapacitors, fuel cells, or micro-turbine generators that complement LIBs were illustrated above for NFCBP buses. The Sinautec America ultracapacitor bus[85] was demonstrated at American University in Washington, D.C., in 2009. Currently, the Massachusetts Institute of Technology spinoff FastCAP Systems is developing nanotechnology hybrid "battacitors" that combine the energy capacity and power density of LIBs and capacitors for future vehicle applications.

In conclusion, the continued federal support from FTA, DOE, and EPA for expanding cleaner and more fuel-efficient bus fleets, in synergy with active ongoing research on improved LIB chemistries and safe operability, promises steady progress in bus fleet electrification and diverse LIB bus applications.[86]

---

[85] See "Sinautec America Demonstrates Bus at America University," http://www.pluginamerica.org/vehicles/sinautec-ultracap-hybrid-bus.

[86] See http://www.fastcapsystems.com/about.html.

ACRONYMS

AC Transit	Alameda Contra Costa Transit
AFCB	American Fuel Cell Bus
AFDC	Alternative Fuels Data Center
ANSI	American National Standards Institute
APTA	American Public Transportation Association
APU	Auxiliary Power Units
ARRA	American Recovery and Reinvestment Act of 2009
AT	Advanced Technology
AVTA	Advanced Vehicle Testing Activities
AZD	Azure Dynamics
BCU	Battery Care Unit
BEB	Battery Electric Bus
BMS	Battery Management System
BYD	Build Your Dream
CARTA	Chattanooga Area Transit Agency
CD	Charge Depleting
CNG	Compressed National Gas
CS	Charge Sustaining
DBNA	Daimler Bus North America
DOE	Department of Energy
DOT	Department of Transportation
EB	Electric Bus
EDSP	Electric Drive Strategy Plan
EMS	Energy Management System
EPA	Environmental Protection Agency
FC	Fuel Cell
FCB	Fuel Cell Bus
FMCSA	National Motor Carrier Safety Administration
FTA	Federal Transit Administration
HEB	Hybrid-Electric Bus
HFC	Hydrogen Fuel Cell
HHICE	Hybrid Hydrogen Internal Combustion Engine
kW	Kilowatt
LCC	Lifecycle Costs

ACRONYMS

LIB	Lithium Ion Battery
LiF3PO4	Lithium Iron Phosphate/Lithium Iron Nanophosphate
LMO	Lithium Magnesium Titanate
LTO	Lithium Ion Titanate
LWFCB	Lightweight Hybrid Fuel Cell Battery
MSDS	Materials Safety Data Sheets
NABI	North American Bus Industries
NaMX	Sodium Metal Halide
NAVC	Northeast Advanced Vehicle Consortium
NCO	Nickel Cobalt Aluminum
NFCBP	National Fuel Cell Bus Program
NFPA	National Fire Protection Association
NHTSA	National Highway Traffic Safety Administration
NiMH	Nickel Metal Hydride
nLTO	Nano-Lithium Ion Titanate
NREL	National Renewable Energy Laboratory
ODI	Office of Defect Investigations
OEM	Original Equipment Manufacturers
PbA	Lead Acid
PEM	Polymer Electrolyte Membrane
PHEB	Plug-in Hybrid-Electric Bus
PM	Permanent Magnet
PMS	Power Management System
R&T	Research and Technology
RESS	Rechargeable Energy Storage System
SAE	Society of Automotive Engineers
SDO	Standard Developing Organizations
SDU	Safety Disconnect Unit
SMFIR	Shaped Magnetic Field-in-Resonance
SOC	State of Charge
TCRP	Transit Cooperative Research Program
TIGGER	Transit Investments for Greenhouse Gas and Energy Reduction
TMS	Thermal Management System
Ucaps	Ultracapacitors

UL	Underwriters Laboratory
USABC	U.S. Advanced Battery Consortium
Volpe Center	Volpe National Transportation Research Center
VTP	Vehicle Technologies Program
ZEB	Zero Emission Buses

www.ingramcontent.com/pod-product-compliance
Lightning Source LLC
Chambersburg PA
CBHW081804170526
45167CB00008B/3320